INSECT WORLD

TERMITES

SANDRA MARKLE

HARDWORKING
INSECT
FAMILIES

◢ LERNER PUBLICATIONS COMPANY MINNEAPOLIS

FOR CURIOUS KIDS EVERYWHERE

ACKNOWLEDGMENTS
The author would like to thank Dr. J. Scott Turner, Biology Department, State University of New York at Syracuse's College of Environmental Science and Forestry, and Dr. Patricia Zungali, Urban Entomology Research Lab, Clemson University, for sharing their expertise and enthusiasm. The author would also like to thank Dr. Simon Pollard, Curator of Invertebrate Zoology at Canterbury Museum, Christchurch, New Zealand, for his help with the scientific name pronunciation guides. Finally, a special thanks to Skip Jeffery, who shared the effort and joy of creating this book.

Lerner Publications Company
A division of Lerner Publishing Group, Inc.
241 First Avenue North
Minneapolis, MN 55401

Website address: www.lernerbooks.com

Library of Congress Cataloging-in-Publication Data

Markle, Sandra.
    Termites : hardworking insect families / by Sandra Markle.
      p.   cm. — (Insect world)
    Includes bibliographical references and index.
    ISBN 978-0-8225-7301-2 (lib. bdg. : alk. paper)   1. Termites—Juvenile literature.
I. Title.
QL529.M375 2008
595.7'36—dc22   2007025963

Manufactured in the United States of America
1 2 3 4 5 6 – DP – 13 12 11 10 09 08

CONTENTS

INSECT WORLD

WELCOME TO THE WORLD OF INSECTS—

those animals nicknamed bugs. It truly is the insects' world. Scientists have discovered more than a million different kinds— more than any other kind of animal. And they are everywhere—even on the frozen continent of Antarctica.

So how can you tell if an animal is an insect rather than a relative such as the scorpion *(right)*? Both termites and scorpions belong to a group of animals called arthropods (AR-throh-podz). The animals in this group share some traits. They have bodies divided into

segments, jointed legs, and a stiff exoskeleton. This is a skeleton on the outside, like a suit of armor. But one sure way to tell if an animal is an insect is to count its legs. All adult insects have six legs. They're the only animals in the world with six legs.

This book is about a kind of insect called the African mound-building termite. These termites live in a colony, or group, that is really one big family. Together they build what are the skyscrapers of the insect world.

TERMITE FACT

Like all insects, a termite's body temperature rises and falls with the temperature around it. So they must warm up to work.

OUTSIDE AND INSIDE

ON THE OUTSIDE

Some people confuse termites and ants. It's easy to tell the difference, though. Compare this termite with the ant. Both have a waist, but the termite's is thicker. Both ants and termites also have antennae. These are the jointed, movable feelers on their heads. But an ant's antennae are usually bent like the bend in an elbow. A termite's antennae are usually straight. They look like strings of beads.

TERMITE FACT

There are over 2,000 different kinds of termites. These can be divided into two main groups: those that live inside wood and those that nest in the ground and go out to find plant material.

TERMITE

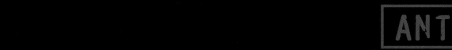

ANT

Take a look at this mound-building termite queen, the member of the family that produces eggs. If you could touch it, its body would feel like tough plastic. Instead of having a hard, bony skeleton on the inside the way you do, an insect has an exoskeleton. This hard coat covers its whole body— even its eyes. The exoskeleton is made up of separate plates connected by stretchy tissue. This lets it bend and move. Check out the other key parts all termites share.

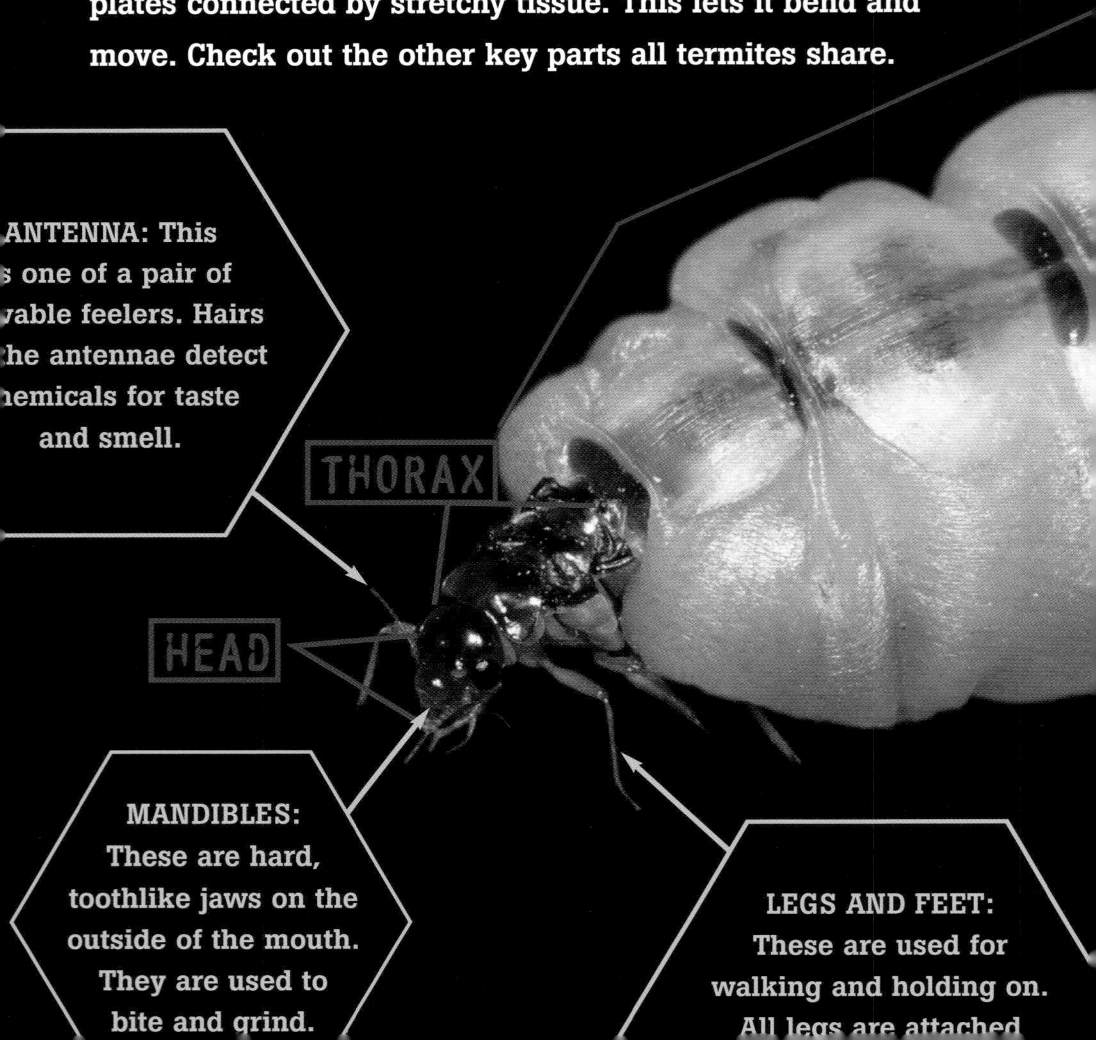

ANTENNA: This is one of a pair of ʋable feelers. Hairs he antennae detect hemicals for taste and smell.

THORAX

HEAD

MANDIBLES: These are hard, toothlike jaws on the outside of the mouth. They are used to bite and grind.

LEGS AND FEET: These are used for walking and holding on. All legs are attached

The termite queen has compound eyes—eyes made up of hundreds of eye units packed together. Termite workers are blind.

ABDOMEN

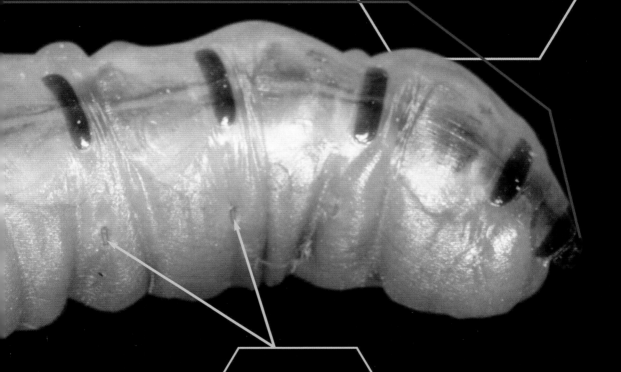

SPIRACLES: These holes down the sides of the thorax and abdomen let air into and out of the body for breathing.

9

ON THE INSIDE

Now look inside the termite queen.

HEART:
This muscular tube pumps blood toward the head. Then the blood flows throughout the body.

ESOPHAGUS:
Food passes through this tube between the mouth and the crop.

BRAIN: This receives messages from the antennae, eyes, and sensory hairs. It sends signals to control all body parts.

CROP: The crop holds food before it moves on for further digestion.

MALPIGHIAN TUBULES: These clean the blood and pass wastes to the intestine.

10

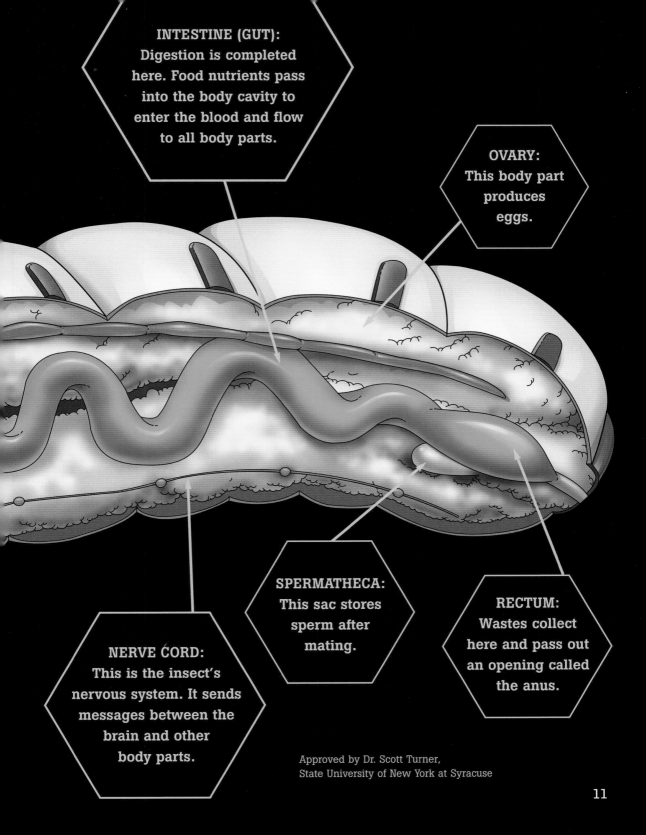

INTESTINE (GUT): Digestion is completed here. Food nutrients pass into the body cavity to enter the blood and flow to all body parts.

OVARY: This body part produces eggs.

SPERMATHECA: This sac stores sperm after mating.

RECTUM: Wastes collect here and pass out an opening called the anus.

NERVE CORD: This is the insect's nervous system. It sends messages between the brain and other body parts.

Approved by Dr. Scott Turner, State University of New York at Syracuse

BECOMING AN ADULT

Insect babies become adults in two ways: incomplete metamorphosis (me-TEH-mor-feh-sus) and complete metamorphosis. Metamorphosis means change. Termites develop through incomplete metamorphosis. Their life includes three stages: egg, nymph, and adult. The nymphs are like the adults in many ways. But nymphs can't reproduce. The nymphs grow into three types of adults: workers, soldiers, and alates (A-lates). Even as adults, workers and soldiers are unable to reproduce. Alates are the adults that go on to create new termite families.

> **IN COMPLETE METAMORPHOSIS,** insects go through four stages: egg, larva, pupa, and adult. Each stage looks and behaves very differently.

NYMPH

EGG

The focus of an African mound-building termite's life is its colony. The colony is really one big family with a single mother and father, called the queen and king. Their offspring become workers, soldiers, and alates. The whole family works together in order to eat and stay safe. They also work together to build and maintain the family's home nest.

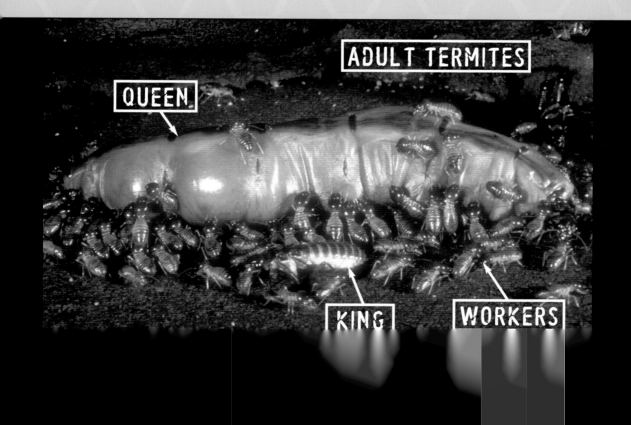

QUEEN

ADULT TERMITES

KING

WORKERS

STARTING A NEW FAMILY

It's a January evening on the African savanna, or grassland. Something is happening deep inside a mound-building termite nest. Hundreds of workers and soldiers are moving through the tunnels. Slowly they guide thousands of winged alates to the surface. Alates are adults able to mate and start new families. It's the first time the alates have been outside the nest. They pause, crowding together. Finally, they take flight.

Some will be caught and eaten by predators. But others will land, pair up, and mate. They'll shed their wings. Then they'll dig into the ground to start building the new mound-building family nest.

TERMITE FACT

Termite families might have hundreds of thousands of members!

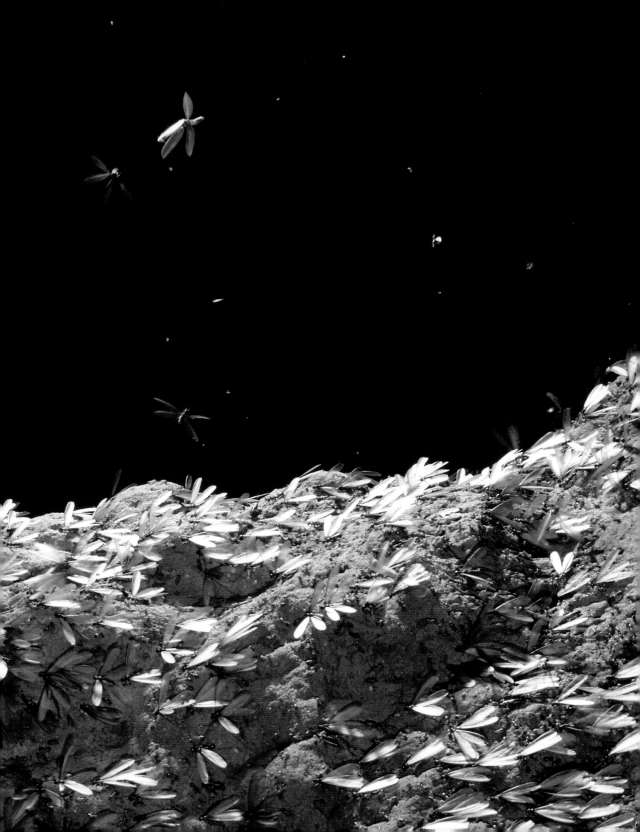

FAMILY TIES

The female lays her first eggs. About three weeks later, the first nymphs hatch. These little males and females are helpless at the beginning. Their mandibles are not yet hard. They can't chew food for themselves. The male and female feed the young by bringing up food stored in their gut. The nymphs grow bigger. Soon their exoskeletons become tight. This covering splits open, and they molt, or shed. Each nymph already has a new exoskeleton underneath. This new covering is soft at first, so the nymphs must wait for it to harden. Then they start eating and growing again. After a second molt, the nymphs have hard mandibles. They are ready to be the family's first workers.

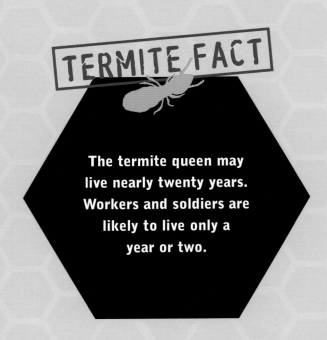

TERMITE FACT

The termite queen may live nearly twenty years. Workers and soldiers are likely to live only a year or two.

The workers take over all the colony's jobs. From then on, the first male and female will be treated like a king and queen. The king's only job will be to mate with the queen. The queen's only job will be to lay eggs. The workers will feed them and clean them. Some workers carry the eggs to nursery chambers. There, other workers feed the young when they hatch. Still other workers start working on the nest, building more tunnels and rooms. They also enlarge the queen's room to make it big enough for her. The more eggs the queen lays, the longer and fatter she becomes.

TERMITE FACT

Termite queens are capable of producing an egg about every three seconds, day and night— as many as 30,000 eggs a day!

SEARCHING AND GARDENING

Another important job for the workers is getting food for the family. Workers leave the nest to find plant material. But they don't go alone. Some nymphs have become soldier termites. They have extra-big mandibles to use as weapons against enemies, like ants. Soldier termites go along with the workers to guard them.

TERMITE FACT

The workers are blind.
They use their senses of smell
and touch to find food.

The mound-building termite workers will eat any kind of plant material they can find. When they come to elephant dung, they dig in. Elephants do not completely digest their food. Some of the food ends up in their dung. The termite workers dig out bits of grass from the dung and swallow them. When their crops are full, they head home again.

TERMITE FACT

As they travel, termite workers touch their abdomen to the ground. Each touch leaves a scent marker. They can then follow this trail home. Other workers follow this trail to the food.

Back at the nest, the workers go to the family's garden room. There they pass their wastes. Other workers pick up these wastes. They use them to build walls, called combs.

Soon fungi start growing on the combs. Fungi are living things similar to yeasts and mushrooms. The workers' wastes have fungi spores, or seedlike parts, in them. As the fungi grow, they break down the plant material in the wastes. The fungi turn the material into food the termites can digest. The workers and soldiers eat this food.

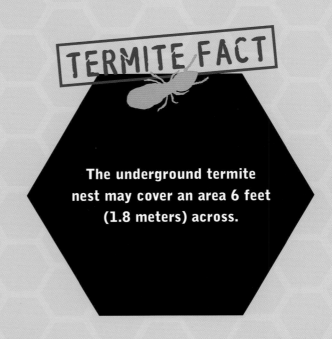

TERMITE FACT

The underground termite nest may cover an area 6 feet (1.8 meters) across.

FAMILY-STYLE DINING

Shortly after eating, the termites pass wastes. Because the wastes were passed so quickly, they still contain food nutrients. These wastes are eaten by other workers and soldiers. Then these workers quickly pass wastes too. Still other termites eat their wastes. This way, the food from the garden is shared by the whole family. Sharing food helps make the termite family a strongly bonded group.

Workers tending the queen not only feed her but they also clean her. This keeps her body free of bacteria and parasites. As they clean the queen, they pick up the special chemicals, called hormones, that she gives off. The workers pass these hormones to other workers through their wastes. Eating the queen's hormones keeps the nymphs from developing into adults that are able to reproduce.

HOME SWEET HOME

When there are lots of workers, the termite family starts building its mound. Bit by bit, the workers carry soil up from underground. At the surface, they mix the soil with their saliva, the juices in their mouth. They stick the soil and saliva mix onto the ground. This way, the workers build a mound that is as hard as cement. The first year, they build a cone-shaped mound. This forms a roof over the underground nest. During the next few years, they add onto it.

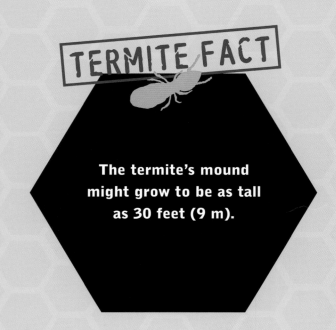

TERMITE FACT

The termite's mound might grow to be as tall as 30 feet (9 m).

The workers build the mound taller and taller. They also build tunnels inside the mound with openings to the outside. When the mound is tall enough, cool breezes enter these openings. Because winds tend to blow harder on one side of the mound than the other, the air flows in on that side. The air flowing around the mound causes a natural suction, which pulls air out of the nest. That keeps fresh air flowing throughout the nest's underground tunnels and rooms. This is important for the termite family to stay healthy. Without a flow of fresh air, the termites would not get enough oxygen to breathe.

TERMITE FACT

A termite nest may outlast a family. It may then become the home of another mound-building termite family.

This yellow–billed hornbill poked its hard bill into the mound. The bird was after a meal of termites. Most of the termites moved out of the way in time. But the hornbill made a hole in the family's home. The termites will have to repair their mound.

As soon as a worker discovers the hole, the worker goes to work. It also gives off special scent chemicals called pheromones (FER-eh-mohnz). This scent attracts other workers. Soon there are lots of termites patching up the hole.

DEFENDING THE FAMILY

One day, ants arrive at the termites' nest. Ants will attack and kill the termites. Then the ants will carry the bodies back to their nest to feed their own colony. The first termite workers the ants meet are killed. But they give off alarm pheromones. This attracts termite soldiers.

The first soldier to arrive attacks the ants with its big, sharp mandibles. It's no match for so many enemies. But soon, more soldiers join the attack. The termites drive the ants away.

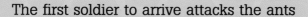

TERMITE FACT

Soldiers with big jaws can't feed themselves. They have to be fed by the other workers.

The mound-building termite family has two kinds of soldiers. They are major soldiers and minor soldiers. Major soldiers, like this one, usually stay inside the nest, guarding the queen and nurseries. They have extra-large heads. They might put their heads together to block a tunnel. Then the enemy can't get through.

MAJOR SOLDIER

Minor soldiers have smaller heads and jaws. They usually go along to guard workers when they search for food.

TERMITE FACT

When fighting enemies, the soldiers produce a drop of brown gluelike substance. This spreads between their jaws. If their bite doesn't kill, the glue may still trap the enemy.

MINOR SOLDIER

THE NEXT GENERATION

Year after year, the queen lays thousands of eggs. Even though some workers and soldiers die, the family keeps getting bigger. Finally, there are so many workers that some get only a very tiny dose of the queen's hormones. These nymphs develop into alates, adult males and females that can reproduce.

If the queen or the king dies, a pair of alates may replace them. More often, the alates swarm out of the nest and fly away. Those that survive start new mound-building termite families.

TERMITE FACT

Termite queens, fried or steamed, are eaten as a special treat in parts of Africa.

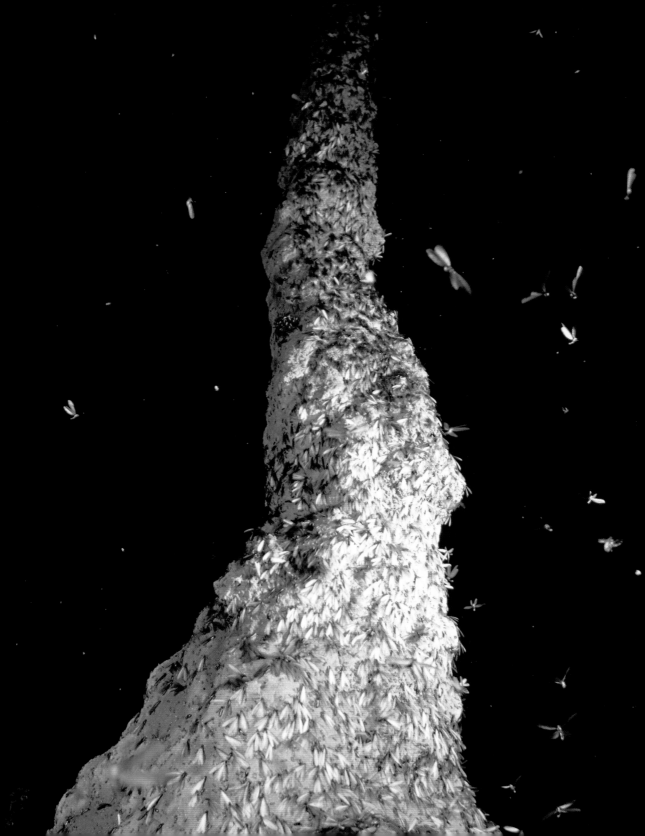

TERMITES AND OTHER INSECT FAMILIES

TERMITES BELONG TO A GROUP, or order, of insects called Isoptera (eye-SOP-ter-ra). That name comes from the Greek words meaning "equal" and "wings." The winged adults have four nearly identical wings. There are about 2,000 kinds of termites in this order. They live in families of a few dozen members to hundreds of thousands.

SCIENTISTS GROUP living and extinct animals with others that are similar. In this system of classification, termites belong to the following groups:

     Kingdom: Animalia
     Phylum: Arthropoda (ar-throh-POH-da)
     Class: Insecta
     Order: Isoptera

HELPFUL OR HARMFUL? Termites are both. Termites are helpful when they break down dead trees and plant material, which would pile up over time. Those that tunnel help loosen hard-packed soil. Termites become pests when they attack the wooden parts of buildings. Termite damage costs as much as $3 billion each year in the United States.

HOW BIG are African mound-building termites? Workers are about 0.19 inches (.5 cm) and queens are about 5 inches (12.7 cm).

WORKER

QUEEN

MORE INSECT FAMILIES

Other insects live and work with their families too. Compare these insect families to the mound-building termites.

Honeybees. The honeybee family, or colony, lives in a hive made of wax cells the bees make. After mating with the female queen, the male dies. The queen lays all the eggs. Honeybees go through complete metamorphosis. The workers build and clean the hive. They feed the larvae. The larvae's food is mainly honey the workers make from nectar, the sweet liquid given off by flowers. The honeybee family lives in its hive for many generations of workers.

Army Ants. The army ant family travels part of the time to search for food. Food is any prey that they can kill or that they find dead. They stop traveling to nest. Then they tunnel underground. The queen ant lays the eggs. Army ants go through complete metamorphosis. The family continues through many generations of workers.

Paper Wasps. Paper wasps go through complete metamorphosis. The paper wasp family starts when the queen begins to build a nest out of paper she makes herself. She lays eggs and raises the first workers. The workers continue adding onto the paper nest. In late summer, the queen produces special eggs. These young grow up to become new queens and males. The old queen and workers die. The family nest is never used again.

GLOSSARY

abdomen: the tail end of an insect. It contains systems for digestion and reproduction.

adult: the final stage of an insect's life cycle. Most insects are able to reproduce at this stage. Termite nymphs develop into three kinds of adults: workers, soldiers, and alates. Only alates are able to reproduce.

alates: winged adult termites that are able to reproduce

antennae: movable, jointed parts on the insect's head used for sensing

brain: receives messages from the antennae, eyes, and sensory hairs. It sends signals to control all body parts.

complete metamorphosis: a process of development in which the young looks and behaves very differently from the adult. Stages include: egg, larva, pupa, and adult.

crop: area of digestive system where food is held before it is passed on for further digestion

egg: a female reproductive cell; also the name given to the first stage of an insect's life cycle

esophagus (ee-SAH-feh-gus)**:** a tube through which food passes from the mouth to the crop, or stomach

exoskeleton: protective, skeleton-like covering on the outside of the body

head: the insect's body part that has the mouth, the brain, and the sensory organs, such as the eyes and the antennae, if there are any

heart: muscular tube that pumps blood

incomplete metamorphosis: a process of development in which the young look and behave much like small adults except that they are unable to reproduce. Stages include: egg, nymph, and adult.

intestine (gut): digestion is completed here. Food nutrients pass into the body cavity to enter the blood and flow to all body parts.

larva: the stage between egg and pupa in complete metamorphosis

Malpighian (mal–PEE-gee-an) **tubules:** the organ that cleans the blood and passes wastes to the intestine

mandibles: the grinding mouthparts of an insect

molt: the process of an insect shedding its exoskeleton

nerve cord: the nervous system. It sends messages between the brain and other body parts.

nymph: stage between egg and adult in incomplete metamorphosis

ovary: body part that produces eggs

pheromones: chemical scents given off as a form of communication

predator: an animal that is a hunter

prey: an animal that a predator catches to eat

pupa: stage between larva and adult in complete metamorphosis. At this stage, the larva's body structure and systems are completely changed into its adult form.

rectum: part of the digestive system where wastes collect before passing out of the body

sperm: male reproductive cell

spermatheca (spur-muh-THEE-kuh)**:** sac in female insects that stores sperm after mating

spiracles (SPIR-i-kehlz)**:** holes down the sides of the thorax and abdomen. They let air into and out of the body for breathing.

thorax: the body part between the head and the abdomen

DIGGING DEEPER

To keep on investigating termites, explore these books and online sites.

BOOKS

Hirschmann, Kris. *Termite.* Farmington Hills, MI: KidHaven Press, 2005. Discover how these insects build their nest.

Schuh, Mari C. *Termites.* Mankato, MN: Pebble Books, 2003. Learn more about how termites live and work.

Telford, Carole, and Rod Theodorou. *Through a Termite City.* Portsmouth, NH: Heinemann, 1998. Explore inside a termite family's home. A map helps you find your way.

WEBSITES

Alien Empire—PBS Online

http://www.pbs.org/wnet/nature/alienempire/metropolis.html

A brief introduction to termites is offered in this PBS website on insects. It includes videos of insect body parts, an insect laying eggs, and a visit to a hive.

Order: Isoptera (Termites)

http://www.museums.org.za/bio/insects/termites/index.htm

A museum in Cape Town, South Africa, runs this site that focuses on mound-building termites.

Totally Termites

http://www.pestworldforkids.org/lesson_termites.html

Discover amazing facts about termites in the United States. Enjoy games and activities.

University of Hawaii Termite Project

http://www.ctahr.hawaii.edu/termite/students/index.htm

Explore termite populations and their life cycles.

TERMITE ACTIVITIES

BUILD LIKE A TERMITE

Termites build their mounds by bringing dirt up to the surface from underground. And they do it one small ball of dirt at a time. Then they stick these balls together using their saliva. Follow the recipe below to prepare salt dough. Then use the dough to build a termite-style mound. As you build, dampen each ball by touching it to a wet sponge. Then stick it to the growing mound.

Cut the top half off a cardboard toilet-paper tube. This will become your mound's center chimney. Use the points of a sharp scissors to poke six holes at different levels in this tube. Stand the tube upright on a sheet of paper. Around the tube's base, form a wall of balls made from salt dough. Before you build the wall higher, cut a plastic straw into six short

chunks. Stick these horizontally into the holes in the tube. Use the scissors points to enlarge the holes if necessary. Add more balls of building material to build up the mound. Mold them around the straw chunks as you go. Trim off any bits of straw that stick out. Build until you reach the top of the tube. Let the mound dry until it's hard.

Salt Dough

Mix together 2 cups of plain flour and 1 cup of table salt. Add 1 cup of water. Stir until well mixed. If the dough is sticky, add more flour. If it's crumbly, add more water. Squeeze the dough in your hands until it's smooth and elastic. To store, wrap with clear wrap and place in the refrigerator.

FEED THE FAMILY RACE

Get the feel for the way a termite family marks and follows a trail to bring home food. You will need at least four people to play. Even more is better. Divide into two teams. Go outdoors. Mark a starting line. Set out objects, like milk jugs, as points both teams will need to go around. Set a finish line. Put a bowl of dried soup beans on this line. Have the teams line up side by side on the starting line.

Give the first person on one team a bag of uncooked elbow macaroni. Give the first person on the other team a bag of popped popcorn. On a starting signal, each team leader will run to the finish line, dropping pasta or popcorn to mark a trail. Team leaders will pick up one bean and follow their own trail back. The next player will then stick to the team's trail to pick up another bean. Have the teams collect beans and carry them home for one minute. When the race is over, the team with the most beans wins. Remember, termite workers mark and follow trails to food. But they are blind. They leave scent markers. The other workers use their antennae to sense and follow this trail to the food. Then they track it to go home again.

INDEX

PHOTO ACKNOWLEDGMENTS

The images in this book are used with the permission of: © Dwight R. Kuhn, p. 4; © Nature's Images/Photo Researchers, Inc., pp. 5, 36; © Piotr Naskrecki/Minden Pictures, p. 7 (top); © Martin Dohrn/naturepl.com, p. 7 (bottom); © Mitsuhiko Imamori/Minden Pictures, pp. 8–9, 17; © Bill Hauser/Independent Picture Service, pp. 10–11; © NHPA/Anthony Bannister, pp. 12, 26–27; © Alan Root/OSF/Animals Animals, p. 13; © Oxford Scientific Films/Photolibrary, pp. 15, 33, 35, 39; © Anthony Bannister; Gallo Images/CORBIS, pp. 19, 34; © Mark Taylor/ naturepl.com, p. 21; © Gregory G. Dimijian, M.D./Photo Researchers, Inc., pp. 22–23; Reprinted with permission of JS Turner. Copyright 2004. All rights reserved, p. 25; © NHPA/Steve Robinson, p. 29; © Simon Williams/naturepl.com, p. 31; © Kim Wolhuter/Gallo Images/Getty Images, p. 32; © Kim Taylor/naturepl.com, pp. 37, 41 (top); © Premaphotos/ naturepl.com, p. 41 (middle); © Kerstin Hinze/naturepl.com, p. 41 (bottom); © Todd Strand/ Independent Picture Service, p. 46.

Front Cover: © Peter Johnson/CORBIS.